National Clean Diesel Rebate Program, 2013 Construction Equipment Funding Opportunity Program Guide

United States
Environmental Protection
Agency

National Clean Diesel Rebate Program, 2013 Construction Equipment Funding Opportunity Program Guide

Transportation and Climate Division
Office of Transportation and Air Quality
U.S. Environmental Protection Agency

United States
Environmental Protection
Agency

EPA-420-B-13-042a
January 2014

Table of Contents

Appendices

1. Introduction

The Environmental Protection Agency (EPA) is offering a 2013 Construction Equipment Funding Opportunity to reduce diesel emissions from existing fleets of nonroad construction equipment. The Diesel Emission Reduction Act program (DERA) was originally authorized by Title VII, Subtitle G (Sections 791 to 797) of the Energy Policy Act of 2005 (Public Law 109-58). DERA was amended by the Diesel Emissions Reduction Act of 2010 (Public Law 111-364), codified at 42 U.S.C. 16131 et seq, adding, among other provisions, a rebate program option. These provisions provide the Environmental Protection Agency with the authority to award rebates, competitive grants and low-cost revolving loans to eligible entities to fund the costs of a clean diesel strategy that significantly reduces diesel emissions from mobile sources through implementation of a certified engine configuration or verified technology. The objective of this program is to achieve significant reductions in diesel emissions in terms of tons of pollution produced and reductions in diesel emissions exposure, particularly from fleets operating in areas designated by the Administrator as poor air quality areas.

1.1 Rebate Program History

The National Clean Diesel Rebate Program was authorized by the Diesel Emissions Reduction Act of 2010. Through the National Clean Diesel Rebate Program, EPA will offer financial support to eligible applicants to reduce diesel emissions from a variety of mobile sources. EPA's first rebate program in 2012 focused on the replacement of school buses.

This second round of rebate funding, known as the 2013 Construction Equipment Funding Opportunity, will provide financial assistance to public and private construction equipment owners for engine replacements or diesel particulate filters on older construction equipment. EPA anticipates offering additional rebate opportunities in future years to retrofit, repower, or replace other types of diesel engines in various sectors. Future rebate funding opportunities will be based on program goals and available funding, among other factors.

1.2 Scope of Work

The 2013 Construction Equipment Funding Opportunity will provide rebate incentives to selected eligible applicants to either: 1) retrofit with a diesel particulate filter or 2) replace their nonroad construction equipment engine. See Section 2.2 for specific information on eligible equipment engines and Section 2.3 for technology options.

The project parameters for the 2013 Construction Equipment Funding Opportunity ensure that all projects that receive funding meet the DERA national priorities. The eligible projects maximize public health benefits, are cost-effective, serve areas that receive a disproportionate quantity of air pollution from diesel fleets, include a certified engine configuration, maximize the useful life of the certified engine configuration, and conserve diesel fuel.

In addition, the 2013 Construction Equipment Funding Opportunity supports EPA's 2011 – 2015 Strategic Plan that defines goals, objectives, and sub-objectives for protecting human health and the environment. Specifically, it supports Goal 1 (Take Action on Climate Change and Improve Air Quality) and Objective 1.2 (Improve Air Quality). Activities funded will reduce diesel emissions from the existing fleet of construction equipment, thereby reducing local and regional air pollution.

This document describes the minimum criteria and requirements of the 2013 Construction Equipment Funding Opportunity.

2. Rebate Program Structure

2.1 Eligible Applicants

Eligible public sector applicants include regional, state, local, or tribal agencies or port authorities with jurisdiction over transportation or air quality. Municipalities, metropolitan planning organizations (MPOs), and counties are all eligible applicants for this rebate program to the extent that they fall within the definition above.

Private entities that operate nonroad construction equipment under a contract or lease with a public entity listed above are also eligible. If the applicant is a private entity, the applicant must certify on the Rebate Application that it has an existing and executed contract or lease to provide nonroad construction equipment to a specified public entity at the time of the rebate application.

2.2 Eligible Construction Equipment Engines

The eligible engines must meet **all** of the definitions and requirements listed below.

Nonroad Engine Definition
The EPA definition of the nonroad engine includes engines installed on: (1) self-propelled equipment; (2) equipment that is propelled while performing its function; or (3) equipment that is portable or transportable, as indicated by the presence of wheels, skids, carrying handles, dolly, trailer, or platform. In other words, nonroad engines are all internal combustion engines except motor vehicle (highway) engines, stationary engines (or engines that remain at one location for more than 12 months), engines used solely for competition, or engines used in aircraft.

Construction Definition
Eligible construction equipment, for the purpose of the 2013 Construction Equipment Funding Opportunity, is defined as nonroad (see above) equipment used in the creation or maintenance of transportation infrastructure, commercial and industrial projects, residential buildings, and heavy civil construction. The construction equipment must be diesel-powered and have a rated power between 130-450 kW or 174-603 horsepower. Examples include but are not limited to wheel or skid loaders, motor graders, and dozers.

Ownership and Annual Usage
At the time of application, the applicant must own and operate the equipment, for which funds are being requested. Equipment ownership must be documented by providing a copy of the original bill of sale, original invoice, or other documentation that demonstrates ownership of the equipment. The equipment must have operated for a minimum of 500 hours in the previous 12 months.

<u>Location Requirement</u>

The nonroad construction equipment must be located in one of the eligible counties in the List of Eligible Counties in Appendix A at the time of application and the 3 months prior to the application. Starting on the date of the rebate payment, following installation of the technology option, the equipment must operate in the application county for 12 months or 500 hours, whichever comes first. The term "eligible county" refers to the primary area where the equipment engines operate, or the primary area where the emissions benefits of the project will be realized. EPA may conduct audits up to 3 years after the rebate payment to ensure these conditions are met.

The counties listed in Appendix A were selected as eligible areas for the 2013 Construction Equipment Funding Opportunity based on data from a number of sources. The sources include counties:

- Designated as PM 2.5 or 8-Hr Ozone Nonattainment Areas or 8-Hr Ozone Maintenance Areas. Data is sourced from EPA's Green Book of Nonattainment Areas for Criteria Pollutants.
  - www.epa.gov/oaqps001/greenbk/
- Where all or part of the population is exposed to more than 2.0 μg/m3 of diesel particulate matter emissions. Data is sourced from the 2005 National-Scale Air Toxics Assessment.
  - www.epa.gov/ttn/atw/nata2005/
- Accepted to participate in EPA's Ozone Advance Program or PM Advance Program by Tuesday, November 19, 2013.
  - www.epa.gov/ozoneadvance/basic.html
  - www.epa.gov/ozoneadvance/basicPM.html

2.3 Technology Options

There are two options under this Funding Opportunity:

1) Retrofit devices for diesel-powered Tier 2 and Tier 3 emission standard nonroad construction engines;
2) Engine replacement for diesel-powered unregulated (Tier 0) and Tier 1 emissions standard nonroad construction engines.

See Table 1 (page 5) and Table 2 (page 6) for specific engine model years and rated power eligibility requirements.

The existing construction equipment must currently operate on diesel fuel, be in regular use, and in operational condition to qualify for funding. To be in regular use, construction equipment engines to be retrofitted or replaced must have accumulated at least 500 annual usage hours over the most recent 12 months. To be considered operational, the equipment must be able to start, have all operational parts in working order, and perform its intended function. If the equipment has more than one engine, all engines are eligible provided the engines on the application meet annual usage hours and rated power requirements.

EPA reserves the right to request maintenance logs or similar at any time during the rebate program process. Applicants are required to certify in the Rebate Application that the construction equipment listed meets these operational requirements.

2.3.1 Technology Option #1 - Engines Eligible for Retrofits with Diesel Particulate Filters

Tier 2 and Tier 3 emission standard nonroad construction engines with engine model years 2001-2010 and power rating of $225 \leq kW < 450$ ($301 \leq hp < 603$) are eligible for rebates on a Diesel Particulate Filter (DPF). EPA will pay for the full cost of the DPF, up to a maximum of $30,000 per DPF.

Please refer to Table 1 and Appendix E to determine eligibility and available rebate amount.

Only verified Diesel Particulate Filters are eligible under this technology option. A list of eligible, EPA verified exhaust control technologies is available at: www.epa.gov/cleandiesel/verification/verif-list.htm; a list of eligible, California Air Resources Board (CARB) verified exhaust control technologies is available at: www.arb.ca.gov/diesel/verdev/vt/cvt.htm. Verified technologies proposed for funding under this category must be specifically named on one of these lists at the time of application to the rebate program, and must be used only for the vehicle application specified on the list.

If selected, applicants will have 45 days to confirm suitability for a DPF by data logging the proposed engine. This can be arranged with the Diesel Particulate Filter vendor. The results of the data logging must be sent to EPA within 45 days of the selection letter.

2.3.2 Technology Option #2 - Engines Eligible for Engine Replacement

Rated Power 130 – 225 kW or 174 - 301 horsepower
Unregulated (Tier 0) and Tier 1 emission standard nonroad construction engines with engine model years 1990-2002 and power rating between $130 – 225$ kW ($174 \leq hp < 301$) are eligible for engine replacements to either Tier 2 or Tier 3 emission standard engines.

Rated Power 225 – 450 kW or 301 - 603 horsepower
Unregulated (Tier 0) and Tier 1 emission standard nonroad construction engines with engine model years 1990-2000 and power rating between $225 – 450$ kW ($301 \leq hp < 603$) are eligible for engine replacements to either Tier 2 or Tier 3 emission standard engines.

Please refer to Table 2 and Appendix F to determine eligibility and available rebate amount.

The Rebate Application must show the pre- and post- project emission standard levels of the engines to be replaced. The replacement engine must be of the same horsepower or within 10% of the engine horsepower being replaced and operate in the same manner as the original engine. All replacement engines must meet Federal safety standards and required warranties. The applicant takes sole responsibility for ensuring the replacement engine is in operational condition.

Following installation of the replacement engine, the original engine must be scrapped by drilling a hole through the engine block and at least one cylinder. See Section 4.5 for specific engine scrappage requirements.

2.4 Available Funding and Selection Process

For the 2013 Construction Equipment Funding Opportunity, EPA anticipates having approximately $2,000,000 for rebates, subject to availability of funds. Funding will not be provided for administrative expenses. EPA reserves the right to partially fund applications, reject all applications, and make no selections under this program, or to make fewer selections than anticipated.

Application selections will be determined by a random lottery. Once the lottery list is generated, EPA will apply the following criteria to the lottery list:

- Each of the 10 EPA Regions, with at least one eligible applicant, will have a selected applicant.
- It is anticipated that at least 50% of funds will be allocated to public construction equipment.

2.5 Amount of Rebate

Tables 1 and 2 specify the rebate amounts for eligible nonroad construction equipment engines.

Table 1: Construction Equipment Rebate Program
Rebate Amount for Technology Option #1

Original Emissions Tier	New Technology	Rated Power 130 ≤ kW < 225 174 ≤ hp < 301	Rated Power 225 ≤ kW < 450 301 ≤ hp < 603
Tier 2 or Tier 3	Diesel Particulate Filter	Not Eligible	$30,000 Max (Eligible Engine Model Years 2001-2010)

Table 2: Construction Equipment Rebate Program
Rebate Amount for Technology Option #2

Original Emissions Tier	New Technology	Rated Power 130 ≤ kW < 225 174 ≤ hp < 301	Rated Power 225 ≤ kW < 450 301 ≤ hp < 603
Unregulated (Tier 0) or Tier 1	Engine Replacement to Tier 2	$12,000 (Eligible Engine Model Years: 1990-2002)	$49,000 (Eligible Engine Model Years: 1990-2000)
Unregulated (Tier 0) or Tier 1	Engine Replacement to Tier 3	$15,000 (Eligible Engine Model Years: 1990-2002)	$69,000 (Eligible Engine Model Years: 1990-2000)

2.6 Maximum Number of Engines per Applicant

Applicants may submit only <u>one</u> Rebate Application that includes up to <u>five</u> eligible nonroad construction equipment engines. The five engines can be any combination of the options shown in the above tables for a total maximum of $120,000 in rebate funding. Applications can include both technology options on the same application, i.e., an applicant could apply for DPF and a replacement engine on the same application.

3. Rebate Process for Technology Option #1 - Retrofit with a Diesel Particulate Filter

The 2013 Construction Equipment Funding Opportunity for Technology Option # 1 consists of the following eight steps, as illustrated in Figure 1. Details of the requirements for each step are described in Sections 3.1 – 3.9 below.

> Step 1 – Application Submission
> Step 2 – Selection of Participants
> Step 3 – Notification of Selectees
> Step 4 – Data Logging Results Submittal
> Step 5 – Purchase Order Submittal
> Step 6 – Delivery and Installation
> Step 7 – Payment Request
> Step 8 – Payment

Figure 1 – Rebate Program Flow Chart for Technology Option # 1 – Retrofit with a Diesel Particulate Filter

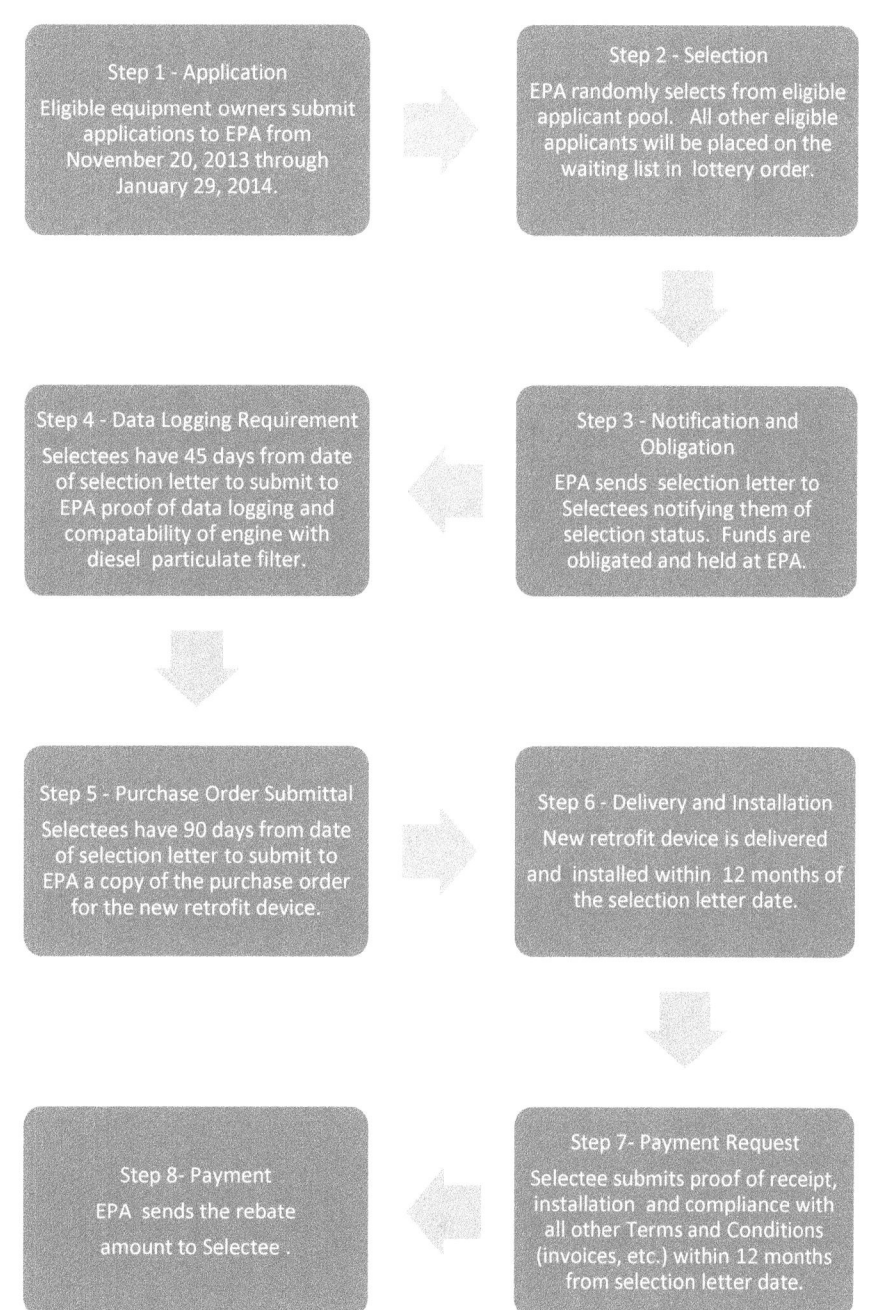

3.1 Step 1 - Application

All applicants must submit a Rebate Application (EPA Form 5600-260) and required supporting documentation to EPA by Wednesday, January 29, 2014, 4:00 pm EST. The application may be downloaded from www.epa.gov/cleandiesel/documents/2013-clean-diesel-rebate-application.pdf as a fillable Portable Document File (PDF). See Appendix B for a sample Rebate Application.

You must have Adobe Reader to open and fill in the fields of this form. For more information about PDFs, please see www.epa.gov/epahome/pdf.html.

Email the completed Rebate Application, along with a scanned copy of the equipment's original bill of sale, invoice or other ownership documentation to CleanDieselRebate@epa.gov. Include the subject line: **DERA Construction Equipment Rebate Application: [your organization's name]**. If the applicant does not have access to email, please call 202-343-9231 for assistance.

The Rebate Application includes identifying information such as organization name, address, Dun and Bradstreet (DUNS) number, Employer ID Number (EIN), and the name of the organization's Authorized Representative, who is able to sign on behalf of the applicant organization. If an applicant does not have a DUNS or EIN, they must obtain one prior to applying for a rebate. An organizational Dun and Bradstreet (D&B) Data Universal Number System (DUNS) number must be included on the rebate form. Organizations may obtain a DUNS number at no cost by calling the toll-free DUNS number request line at 1-866-705-5711, or visiting the D&B website at: www.dnb.com.

Applicants must also be registered in the System for Award Management (SAM), prior to submitting an application (previously known as the Central Contractor Registration). Information can be found at www.sam.gov.

The Rebate Application requires applicants to supply the following information related to each of the construction equipment engines to be retrofitted.

1) Equipment Type
2) Technology Option and Emission Standard Tier
3) Engine Serial Number (see Appendix D)
4) Engine Family Name (see Appendix D)
5) Engine Model Year (see Appendix D)
6) Gallons of Diesel Used in Last 12 Months
7) Hours of Use Last 12 Months
8) Horsepower
9) Location of operation of the construction equipment (County and State)
10) Rebate Amount (See Section 2.5 and Appendix E for eligible rebate amounts)

The engine model year of the existing equipment's engine must be between 2001 and 2010, as described in Section 2.3.1 and Table 1. See Appendix D for additional assistance with determining the engine serial number, engine model year, or engine family name.

A copy of the equipment's original bill of sale, invoice, or other documentation that proves ownership must be submitted with the application. See Section 2.2 for additional information.

Annual usage hours should be obtained from maintenance logs or other recordkeeping information. By signing the Rebate Application, applicants are certifying that the engine serial number, engine model year, engine family name, horsepower, annual usage hours, and fuel usage reported are true to the best of their knowledge. EPA reserves the right to request copies of documentation, such as activity logs, to verify the above information.

EPA may contact an applicant to clarify any information provided by that applicant.

3.2 Step 2 - Selection

All applications that are received by EPA by **Wednesday, January 29, 2014, 4:00 pm EST,** will be assigned a unique identification number, and applicants will be selected through a random number generator. Once the lottery list is generated, EPA will apply the following criteria to the lottery list:

- Each of the 10 EPA Regions, with at least one eligible applicant, will have a selected applicant.
- It is anticipated that at least 50% of funds will be allocated to public construction equipment.

Eligible applicants who are selected (Selectees) will move on to Step 3 in the Rebate Process. All other applicants will remain in lottery number order on the wait list. If a Selectee does not complete the remaining required steps in the rebate process within the required timeframe (described in Steps 3 – 8 below), that Selectee will be removed from the program and the next applicant on the wait list will be selected for participation.

Both the Selectee List and the Applicant Wait List will be posted at www.epa.gov/cleandiesel/dera-rebate-construction.htm

3.3 Step 3 – Notification and Obligation

EPA will notify Selectees and those applicants that are on the wait list within 30 days of the application submittal deadline. Each Selectee will be assigned an EPA Rebate Contact person for the duration of the Rebate Program. Once Selectees receive the selection letter from EPA, they may purchase the eligible DPFs and complete Steps 4 – 8 below.

3.4 Step 4 – Data Logging Results Submittal

The Selectee will have 45 days to confirm suitability for a DPF by data logging the proposed engine for two weeks. This can be arranged with the Diesel Particulate Filter vendor. The Selectee must submit the results of the data logging process to their Rebate Contact within 45 days of the Selection Letter.

3.5 Step 5 – Purchase Order Submittal

A copy of the purchase order for the new DPF must be submitted to EPA within 90 days of date of the Selection Letter. The date of the purchase order cannot pre-date the date of the Selection Letter. The proof of purchase may be a procurement request, purchase order, or any other document that clearly shows a transaction being initiated between the applicant and a retrofit vendor for the purchase of an eligible DPF.

Proof of purchase must be on official vendor or purchaser letterhead and include the following information for each piece of equipment: (1) purchaser name, address, and phone number; (2) vendor name, address, and phone number; (3) diesel particulate filter model, manufacturer and purchase price; (4) diesel particulate filter purchase date; (5) delivery date(s), predetermined and agreed upon by both vendor and purchaser. Selectees that submit false or misleading information may be barred from future participation in DERA and other federal funding programs or may face other penalties.

The proof of purchase document should be scanned and saved in PDF format and emailed to: your assigned EPA Rebate Contact and CleanDieselRebate@epa.gov. Include the subject line: **DERA Construction Equipment Proof of Purchase: [your organization's name]**.

3.6 Step 6 – Diesel Particulate Filter Delivery and Installation
Selectees must take delivery and install the new DPF within 12 months of the date of the Selection Letter and prior to submitting the Payment Request to EPA. New DPFs and retrofitted engines must meet the requirements described in Section 2.3.1.

3.7 Step 7 - Request for Payment
Selectees may request reimbursement from EPA by submitting a Payment Request, a copy of the invoice, and a copy of the bill of lading (proof of delivery) for the new DPF. The invoice or the bill of lading should include the following information: Manufacturer and Model of the DPF, Engine Model Year, Engine Manufacturer, Engine Family Name, and DPF cost. The Payment Request Form also requires Selectees to provide detailed information on the retrofitted engine and DPF, such as the Manufacturer and Model of the DPF, Engine Model Year, Engine Manufacturer, Engine Family Name, and DPF cost. The Payment Request must be submitted to EPA no later than 12 months after the date of the Selection Letter (see Section 3.3 for information about the Selection Letter).

To request reimbursement, email the completed Payment Request, a copy of the DPF invoice, and a copy of the bill of lading (in PDF format) as attachments to your assigned EPA Rebate Contact and CleanDieselRebate@epa.gov. Include the subject line: **DERA Construction Equipment Payment Request: [your organization's name]**.

3.8 Step 8 - Payment
Once EPA has received and approved the Selectee's Payment Request and supporting documentation, EPA will issue the rebate funds electronically to the Selectee. EPA anticipates that payment will be issued to the Selectee within approximately 10 business days from the receipt of the complete payment request package.

3.9 Cancellation of Rebate Application
If a Selectee fails to submit all of the required forms and other documents by the deadlines established in Section 3.4, 3.5, and 3.7, the rebate application will be cancelled and any reserved funds will be offered to the next eligible applicant on the waiting list. EPA will notify the Selectee prior to cancelling any reserved funds.

4. Rebate Process for Technology Option # 2 - Replacement Engine to Tier 2 or Tier 3 Emission Standard Engine

The 2013 Construction Equipment Funding Opportunity consists of the following seven steps, as illustrated in Figure 2. Details of the requirements for each step are described in Sections 4.1 – 4.8 below.

Step 1 – Application Submission
Step 2 – Selection of Participants
Step 3 – Notification of Selectees
Step 4 – Purchase Order Submittal
Step 5 – Delivery and Scrappage
Step 6 – Payment Request
Step 7 – Payment

Figure 2 – Rebate Program Flow Chart for Technology Option # 2 – Replacement Engine to Tier 2 or Tier 3 Emission Standard Engine

4.1 Step 1 - Application

All applicants must submit a Rebate Application (EPA Form 5600-260) and required supporting documentation to EPA by Wednesday, January 29, 2014, 4:00 pm EST. The application may be downloaded from www.epa.gov/cleandiesel/documents/2013-clean-diesel-rebate-application.pdf as a fillable Portable Document File (PDF). See Appendix B for a sample Rebate Application.

You must have Adobe Reader to open and fill in the fields of this form. For more information about PDFs, please see www.epa.gov/epahome/pdf.html.

Email the completed Rebate Application, along with a scanned copy of the equipment's original bill of sale, invoice or other ownership documentation to: CleanDieselRebate@epa.gov. Include the subject line: **DERA Construction Equipment Rebate Application: [your organization's name]**. If the applicant does not have access to email, please call 202-343-9231 for assistance.

The Rebate Application includes identifying information such as organization name, address, Dun and Bradstreet (DUNS) number, Employer ID Number (EIN), and the name of the organization's Authorized Representative, who is able to sign on behalf of the applicant organization. If an applicant does not have a DUNS or EIN, they must obtain one prior to applying for a rebate. An organizational Dun and Bradstreet (D&B) Data Universal Number System (DUNS) number must be included on the rebate form. Organizations may obtain a DUNS number at no cost by calling the toll-free DUNS number request line at 1-866-705-5711, or visiting the D&B website at: www.dnb.com.

Applicants must also be registered in the System for Award Management (SAM), prior to submitting an application (previously known as the Central Contractor Registration). Information can be found at www.sam.gov.

The Rebate Application requires applicants to supply the following information related to the construction equipment engine to be replaced.

1) Equipment Type
2) Technology Option and Emission Standard Tier
3) Engine Serial Number (see Appendix D)
4) Engine Family Name (see Appendix D)
5) Engine Model Year (see Appendix D)
6) Gallons of Diesel Used in Last 12 Months
7) Hours of Use Last 12 Months
8) Horsepower
9) Location of operation of the construction equipment (County and State)
10) Rebate Amount (See Section 2.5 and Appendix F for eligible rebate amounts)

The engine model year of the existing equipment's engine must be between 1990 and 2002 for Unregulated (Tier 0) emission standard engines or between 1990 and 2000 for Tier 1 emission standard engines, as described in Section 2.3.2 and Table 2. See Appendix D for additional assistance with determining the engine serial number, engine model year, or engine family name.

A copy of the equipment's original bill of sale, invoice, or other documentation that proves ownership must be submitted with the application. See Section 2.2 for additional information.

Annual usage hours should be obtained from maintenance logs or other recordkeeping information. By signing the Rebate Application, applicants are certifying that the engine serial number, engine model year, engine family name, horsepower, annual usage hours, and fuel usage reported are true to the best of their knowledge. EPA reserves the right to request copies of documentation, such as activity logs, to verify the above information.

EPA may contact an applicant to clarify any information provided by that applicant.

4.2 Step 2 - Selection

All applications that are submitted to EPA by **Wednesday, January 29, 2014, 4:00 pm EST,** will be assigned a unique identification number, and applicants will be selected through a random number generator. Once the lottery list is generated, EPA will apply the following criteria to the lottery list:

- Each of the 10 EPA Regions, with at least one eligible applicant, will have a selected applicant.
- It is anticipated that at least 50% of funds will be allocated to public construction equipment.

Eligible applicants who are selected (Selectees) will move on to Step 3 in the Rebate Process. All other applicants will remain in random number order on the wait list. If a Selectee does not complete the remaining required steps in the rebate process within the required timeframe (described in Steps 3 – 7 below), that Selectee will be removed from the program and the next applicant on the wait list will be selected for participation.

Both the Selectee List and the Applicant Wait List will be posted at www.epa.gov/cleandiesel/dera-rebate-construction.htm

4.3 Step 3 – Notification and Obligation

EPA will notify Selectees and those applicants that are on the wait list within 30 days of the application submittal deadline. Each Selectee will be assigned an EPA Rebate Contact person for the duration of the Rebate Program. Once Selectees receive the selection letter from EPA, they may begin the engine replacement process. Once the engine has been ordered, the Selectee will complete Steps 4 – 7 below.

4.4 Step 4 - Purchase Order Submittal

A copy of the purchase order(s) for the new replacement engine must be submitted to EPA within 90 days of date of the selection letter. The date of the purchase order cannot pre-date the date of the Selection Letter. The proof of purchase may be a procurement request, purchase order, or any other document that clearly shows a transaction being initiated between the applicant and an engine vendor for the purchase of an eligible replacement engine.

Proof of purchase must be on official vendor or purchaser letterhead and include the following information for each engine: (1) purchaser name, address, and phone number; (2) vendor name, address, and phone number; (3) engine serial number, engine manufacturer, engine model year, emission standard tier, engine family and engine purchase price; (4) replacement engine purchase date; (5) delivery date, predetermined and agreed upon by both vendor and purchaser. Selectees that submit false or misleading information may be barred from future participation in DERA and other federal funding programs or may face other penalties.

The proof of purchase document should be scanned and saved in PDF format and emailed to: your assigned EPA Rebate Contact and CleanDieselRebate@epa.gov. Include the subject line: **DERA Construction Equipment Proof of Purchase: [your organization's name]**.

4.5 Step 5 – Vehicle Delivery and Scrappage

4.5.1 – New Engine Delivery and Installation
Within 12 months of the Selection Letter date, Selectees must take delivery of and install the new engine prior to submitting the Payment Request to EPA. New engines must meet the requirements described in Section 2.3.2.

4.5.2 Scrappage/Disposal of Old Engine
Each engine being replaced must be scrapped or rendered permanently disabled prior to Selectee submitting the Payment Request to EPA. The older engine being replaced must be scrapped and permanently disabled by drilling a minimum ½" diameter hole completely through the engine block and at least one cylinder and cutting through the intake manifold.

Proof of scrappage must be provided with the Payment Request. Scrappage documentation includes photos of:
1. Side profile of the equipment
2. The engine tag that includes:
    a. Engine serial number
    b. Engine family identifier
3. Engine block, prior to hole being drilled
4. Engine block, after hole has been drilled

The Selectee must also provide a letter confirming the scrappage requirements have been met. The letter must be signed by the authorized representative listed on the rebate application forms. The letter should include:
1. The date the engines were scrapped.
2. A listing of the scrapped engines with engine model year, engine serial number, horsepower and emission standard tier level.
3. The name and contact information for the entity that scrapped the equipment, if other than the applicant.

All scrappage documentation should be submitted, with the Payment Request, in PDF format or as JPEG image files.

Scrappage may be completed by the Selectee or by a salvage yard, or similar service, provided all scrappage requirements have been met and all necessary documentation is provided. The engine may be sold for scrap metal, provided that the engine is disposed of in accordance with federal and state requirements for disposal.

4.6 Step 6 - Request for Payment
Selectees may request reimbursement from EPA by submitting a Payment Request, proof of scrappage (see Section 4.5.2), a copy of the invoice for the new engine, and a copy of the bill of

lading (proof of delivery) for the new engine. The invoice or bill of lading for the replacement engine should include the following information: Engine Serial Number, Engine Model Year, Engine Manufacturer, Engine Family Name, Emission Standard Tier, Horsepower and Engine Cost.

The Payment Request requires Selectees to provide detailed information on the new engine, such as the Engine Serial Number, Engine Model Year, Engine Manufacturer, Engine Family Name, Emission Standard Tier, Horsepower and Cost. The Payment Request must be submitted to EPA no later than 12 months after the date of the Selection Letter (see Section 4.3 for information about the Selection Letter).

To request reimbursement, email the completed Payment Request, proof of scrappage, a copy of the replacement engine invoice, and a copy of the bill of lading (in PDF format) as attachments to: your assigned EPA Rebate Contract and CleanDieselRebate@epa.gov. Include the subject line: **DERA Construction Equipment Payment Request: [your organization's name]**.

4.7 Step 7 - Payment

Once EPA has received and approved the Selectee's Payment Request and supporting documentation, EPA will issue the rebate funds electronically to the Selectee. EPA anticipates that payment will be issued to the Selectee within approximately 10 business days from the receipt of the complete payment request package.

4.8 Cancellation of Rebate Application

If a Selectee fails to submit all of the required forms and other documents by the deadlines established in Sections 4.4 and 4.6, the rebate application will be cancelled and any reserved funds will be offered to the next eligible applicant on the waiting list. EPA will notify the Selectee prior to cancelling any reserved funds.

5. Rebate Program Administration

5.1 Terms and Conditions

Applicants are required to comply with the following terms and conditions. By signing the Rebate Application (see Appendix B for example), applicants certify that they have read and agree to the requirements of this Program Guide document and the program terms and conditions.

5.1.1 Use of Construction Equipment Engine and/or Retrofits

The equipment engine must operate in a similar manner as the engine prior to retrofit or replacement. In addition, the Selectee agrees that they will:

1. Not make modifications to the emission control system on the replacement engine or the retrofit device; and,
2. Be available for follow-up inspection of the piece of equipment for 3 years after receipt of the rebate, if requested by EPA or its designee. EPA anticipates auditing of a random sample of rebate recipients.

5.1.2 Scrappage of Old Equipment Engine for Replaced Engines Only

Selectees must scrap the engine being replaced in accordance with Section 4.5.2 (Scrappage/Disposal of Old Engine).

5.1.3 Ownership and Location of the Retrofitted or Replaced Engines

Selectees must maintain ownership of the retrofitted or replaced engine for 12 months following payment of the rebate. If the equipment is sold or moved outside of the priority county before the end of the 12 month period or 500 hours of operation, whichever comes first, the Selectee may be required to return up to the full amount of the rebate to EPA. The amount required to be returned is at the discretion of EPA, and will be determined on a case-by-case basis.

5.1.4 Restriction for Mandated Measures

Pursuant to 42 U.S.C. 16132(d)(2), no funds awarded under the 2013 Construction Equipment Funding Opportunity shall be used to fund the costs of emission reductions that are mandated under federal law. In addition, these federal funds for construction equipment engines must not be used in combination with any other federal funding.

5.1.5 Documentation Requirement

If an applicant is selected, that Selectee is responsible for providing the Payment Request and required supporting documentation to EPA. Selectees are responsible for maintaining copies of all submitted forms and documents, and EPA responses, for a period of 3 years from the date of payment.

5.2 EPA Responsibilities

EPA will review rebate applications for eligibility within the timelines established in Section 3.3 and 4.3. EPA will promptly notify Selectees by email and post the waiting list to our website within 30 days of the close of the open application period.

5.3 Disbursement of Funds

EPA will issue rebate funds within 10 business days of determining that a Selectee has submitted a completed Payment Request and all supporting documents, including proof of scrappage of the old engine (for engine replacements only) and proof of the new technology installation and delivery. If necessary, EPA may request additional documentation from a Selectee prior to issuing funds if EPA determines that any required information is missing or incomplete. In such a case, EPA will provide the Selectee with a reasonable amount of time to submit additional information.

5.4 Emission Reductions Reporting

EPA will use the equipment information supplied by applicants to calculate emissions reductions attributable to the 2013 Construction Equipment Funding Opportunity for the purposes of program evaluation and reporting to Congress on the effectiveness of the program.

5.5 Program Audit

EPA will conduct random reviews of Selectees' documentation to protect against waste, fraud, and abuse. As part of this process, EPA may request copies of rebate documents from prior Selectees who have received rebates, or may request documentation from current Selectees to verify statements made on the application and payment forms. Selectees are expected to comply with recordkeeping requirements (see Section 5.1.5), and must supply EPA with any requested documents for 3 years from date of rebate issuance, or risk cancellation of an active rebate application or other enforcement action.

5.6   Record Retention Requirements

Selectees must retain all financial records, supporting documents, accounting books, and other evidence of Rebate Program activities for 3 years. The retention period starts on the day the Applicant is notified that their application has been selected for funding.   If any litigation, claim, or audit is started before the expiration of the three year period, the recipient must maintain all appropriate records until these actions are completed and all issues resolved.

Appendix A
List of Priority Counties Eligible for 2013 Construction Equipment Funding Opportunity

AL	Jefferson	CA	Tehama	GA	Douglas
AL	Mobile	CA	Tulare	GA	Fayette
AL	Shelby	CA	Ventura	GA	Forsyth
AL	Walker	CA	Washoe Tribal Lands in CA	GA	Fulton
AK	Anchorage			GA	Gwinnett
AK	Fairbanks North Star	CA	Yolo	GA	Henry
AZ	Maricopa	CA	Yuba	GA	Houston
AZ	Pima	CO	Adams	GA	Jones
AZ	Pinal	CO	Arapahoe	GA	Monroe
AZ	Santa Cruz	CO	Boulder	GA	Muscogee
AR	Crittenden	CO	Denver	GA	Newton
AR	Pulaski	CO	Douglas	GA	Paulding
CA	Alameda	CO	Jefferson	GA	Peach
CA	Butte	CO	Larimer	GA	Rockdale
CA	Calaveras	CO	Weld	GA	Twiggs
CA	Contra Costa	CT	Fairfield	HI	Honolulu
CA	El Dorado	CT	Hartford	ID	Ada
CA	Fresno	CT	Litchfield	ID	Franklin
CA	Imperial	CT	Middlesex	ID	Twin Falls
CA	Kern	CT	New Haven	IL	Cook
CA	Kings	CT	New London	IL	DuPage
CA	Los Angeles	CT	Tolland	IL	Grundy
CA	Madera	CT	Windham	IL	Kane
CA	Marin	DE	Kent	IL	Kendall
CA	Mariposa	DE	New Castle	IL	Lake
CA	Merced	DE	Sussex	IL	McHenry
CA	Napa	DC	District of Columbia	IL	Madison
CA	Nevada	FL	Brevard	IL	Monroe
CA	Orange	FL	Broward	IL	Rock Island
CA	Placer	FL	Duval	IL	St. Clair
CA	Riverside	FL	Hillsborough	IL	Will
CA	Sacramento	FL	Lee	IL	Winnebago
CA	San Bernardino	FL	Miami-Dade	IN	Clark
CA	San Diego	FL	Orange	IN	Dearborn
CA	San Francisco	FL	Palm Beach	IN	Lake
CA	San Joaquin	FL	Sarasota	IN	Marion
CA	San Luis Obispo	GA	Bartow	IN	Porter
CA	San Mateo	GA	Bibb	IA	Black Hawk
CA	Santa Barbara	GA	Chatham	IA	Harrison
CA	Santa Clara	GA	Cherokee	IA	Johnson
CA	Shasta	GA	Clayton	IA	Linn
CA	Solano	GA	Cobb	IA	Mills
CA	Sonoma	GA	Coweta	IA	Polk
CA	Stanislaus	GA	Crawford	IA	Pottawattamie
CA	Sutter	GA	DeKalb	IA	Scott

IA	Woodbury		MD	Baltimore City		MN	Kandiyohi
KS	Butler		MA	Bristol		MN	Kittson
KS	Harvey		MA	Dukes		MN	Koochiching
KS	Johnson		MA	Middlesex		MN	Lac qui Parle
KS	Leavenworth		MA	Suffolk		MN	Lake
KS	Miami		MI	Genesee		MN	Lake of the Woods
KS	Sedgwick		MI	Ingham		MN	Le Sueur
KS	Sumner		MI	Kent		MN	Lincoln
KS	Wyandotte		MI	Livingston		MN	Lyon
KY	Boone		MI	Macomb		MN	McLeod
KY	Campbell		MI	Monroe		MN	Mahnomen
KY	Fayette		MI	Oakland		MN	Marshall
KY	Jefferson		MI	St. Clair		MN	Martin
KY	Kenton		MI	Washtenaw		MN	Meeker
LA	Ascension		MI	Wayne		MN	Mille Lacs
LA	Assumption		MN	Aitkin		MN	Morrison
LA	Bossier		MN	Anoka		MN	Mower
LA	Caddo		MN	Becker		MN	Murray
LA	Calcasieu		MN	Beltrami		MN	Nicollet
LA	Cameron		MN	Benton		MN	Nobles
LA	De Soto		MN	Big Stone		MN	Norman
LA	East Baton Rouge		MN	Blue Earth		MN	Olmsted
LA	Iberville		MN	Brown		MN	Otter Tail
LA	Jefferson		MN	Carlton		MN	Pennington
LA	Lafourche		MN	Carver		MN	Pine
LA	Lafayette		MN	Cass		MN	Pipestone
LA	Livingston		MN	Chippewa		MN	Polk
LA	Orleans		MN	Chisago		MN	Pope
LA	Plaquemines		MN	Clay		MN	Ramsey
LA	St. Bernard		MN	Clearwater		MN	Red Lake
LA	St. Charles		MN	Cook		MN	Redwood
LA	St. James		MN	Cottonwood		MN	Renville
LA	St. John the Baptist		MN	Crow Wing		MN	Rice
LA	St. Tammany		MN	Dakota		MN	Rock
LA	Terrebonne		MN	Dodge		MN	Roseau
LA	West Baton Rouge		MN	Douglas		MN	St. Louis
ME	Cumberland		MN	Faribault		MN	Scott
MD	Anne Arundel		MN	Fillmore		MN	Sherburne
MD	Baltimore		MN	Freeborn		MN	Sibley
MD	Calvert		MN	Goodhue		MN	Stearns
MD	Carroll		MN	Grant		MN	Steele
MD	Cecil		MN	Hennepin		MN	Stevens
MD	Charles		MN	Houston		MN	Swift
MD	Frederick		MN	Hubbard		MN	Todd
MD	Harford		MN	Isanti		MN	Traverse
MD	Howard		MN	Itasca		MN	Wabasha
MD	Montgomery		MN	Jackson		MN	Wadena
MD	Prince George's		MN	Kanabec		MN	Waseca

MN	Washington		NE	Douglas		NC	Mecklenburg
MN	Watonwan		NE	Lancaster		NC	Rowan
MN	Wilkin		NE	Sarpy		NC	Union
MN	Winona		NE	Saunders		OH	Ashtabula
MN	Wright		NE	Washington		OH	Butler
MN	Yellow Medicine		NV	Clark		OH	Clark
MS	DeSoto		NV	Washoe Tribal Lands in NV		OH	Clermont
MS	Hancock		NJ	Atlantic		OH	Clinton
MS	Harrison		NJ	Bergen		OH	Cuyahoga
MS	Hinds		NJ	Burlington		OH	Darke
MS	Jackson		NJ	Camden		OH	Delaware
MS	Warren		NJ	Cape May		OH	Fairfield
MS	Washington		NJ	Cumberland		OH	Franklin
MO	Barry		NJ	Essex		OH	Geauga
MO	Barton		NJ	Gloucester		OH	Greene
MO	Bollinger		NJ	Hudson		OH	Hamilton
MO	Cape Girardeau		NJ	Hunterdon		OH	Jefferson
MO	Cass		NJ	Mercer		OH	Knox
MO	Cedar		NJ	Middlesex		OH	Lake
MO	Christian		NJ	Monmouth		OH	Licking
MO	Clay		NJ	Morris		OH	Lorain
MO	Dade		NJ	Ocean		OH	Lucas
MO	Dallas		NJ	Passaic		OH	Madison
MO	Franklin		NJ	Salem		OH	Medina
MO	Greene		NJ	Somerset		OH	Miami
MO	Iron		NJ	Sussex		OH	Montgomery
MO	Jackson		NJ	Union		OH	Portage
MO	Jasper		NJ	Warren		OH	Preble
MO	Jefferson		NM	Bernalillo		OH	Stark
MO	Lawrence		NY	Albany		OH	Summit
MO	McDonald		NY	Bronx		OH	Warren
MO	Madison		NY	Chautauqua		OK	Canadian
MO	Newton		NY	Kings		OK	Cleveland
MO	Perry		NY	Monroe		OK	Creek
MO	Platte		NY	Nassau		OK	Grady
MO	Polk		NY	New York		OK	Lincoln
MO	Ray		NY	Onondaga		OK	Logan
MO	St. Charles		NY	Orange		OK	McClain
MO	Ste. Genevieve		NY	Queens		OK	Oklahoma
MO	St. Francois		NY	Richmond		OK	Okmulgee
MO	St. Louis		NY	Rockland		OK	Osage
MO	Stone		NY	Suffolk		OK	Pawnee
MO	Taney		NY	Westchester		OK	Rogers
MO	Webster		NC	Cabarrus		OK	Tulsa
MO	St. Louis City		NC	Cumberland		OK	Wagoner
MT	Lewis and Clark		NC	Gaston		OR	Clackamas
MT	Silver Bow		NC	Iredell		OR	Klamath
NE	Cass		NC	Lincoln		OR	Lake

OR	Lane	SC	Florence	TX	El Paso
OR	Multnomah	SC	Georgetown	TX	Falls
OR	Washington	SC	Greenville	TX	Fort Bend
PA	Allegheny	SC	Greenwood	TX	Freestone
PA	Armstrong	SC	Hampton	TX	Galveston
PA	Beaver	SC	Horry	TX	Gregg
PA	Berks	SC	Jasper	TX	Guadalupe
PA	Bucks	SC	Kershaw	TX	Harris
PA	Butler	SC	Lancaster	TX	Harrison
PA	Cambria	SC	Laurens	TX	Hays
PA	Carbon	SC	Lee	TX	Hill
PA	Chester	SC	Lexington	TX	Hood
PA	Cumberland	SC	McCormick	TX	Jefferson
PA	Dauphin	SC	Marion	TX	Johnson
PA	Delaware	SC	Marlboro	TX	Kaufman
PA	Fayette	SC	Newberry	TX	Kendall
PA	Greene	SC	Oconee	TX	Liberty
PA	Indiana	SC	Orangeburg	TX	Limestone
PA	Lancaster	SC	Pickens	TX	Lubbock
PA	Lawrence	SC	Richland	TX	McLennan
PA	Lebanon	SC	Saluda	TX	Montgomery
PA	Lehigh	SC	Spartanburg	TX	Nueces
PA	Montgomery	SC	Sumter	TX	Parker
PA	Northampton	SC	Union	TX	Potter
PA	Philadelphia	SC	Williamsburg	TX	Randall
PA	Washington	SC	York	TX	Rockwall
PA	Westmoreland	TN	Anderson	TX	Rusk
PA	York	TN	Blount	TX	San Patricio
SC	Abbeville	TN	Davidson	TX	Smith
SC	Aiken	TN	Hamilton	TX	Tarrant
SC	Allendale	TN	Knox	TX	Taylor
SC	Anderson	TN	Loudon	TX	Travis
SC	Bamberg	TN	Roane	TX	Upshur
SC	Barnwell	TN	Shelby	TX	Waller
SC	Beaufort	TX	Atascosa	TX	Webb
SC	Berkeley	TX	Bastrop	TX	Wichita
SC	Calhoun	TX	Bexar	TX	Williamson
SC	Charleston	TX	Bosque	TX	Wilson
SC	Cherokee	TX	Bowie	TX	Wise
SC	Chester	TX	Brazoria	UT	Box Elder
SC	Chesterfield	TX	Brazos	UT	Cache
SC	Clarendon	TX	Caldwell	UT	Davis
SC	Colleton	TX	Chambers	UT	Duchesne
SC	Darlington	TX	Collin	UT	Salt Lake
SC	Dillon	TX	Comal	UT	Tooele
SC	Dorchester	TX	Dallas	UT	Uintah
SC	Edgefield	TX	Denton	UT	Uintah and Ouray Reservation
SC	Fairfield	TX	Ellis	UT	Utah

UT	Weber	VA	Colonial Heights	WA	Snohomish
VA	Arlington	VA	Fairfax City	WA	Yakima
VA	Caroline	VA	Falls Church	WV	Brooke
VA	Charles City	VA	Hampton	WV	Hancock
VA	Chesterfield	VA	Hopewell	WV	Kanawha
VA	Fairfax	VA	Manassas	WV	Putnam
VA	Gloucester	VA	Manassas Park	WV	Wayne
VA	Hanover	VA	Newport News	WI	Brown
VA	Henrico	VA	Norfolk	WI	Dane
VA	Isle of Wight	VA	Petersburg	WI	Kenosha
VA	James City	VA	Poquoson	WI	Milwaukee
VA	Loudoun	VA	Portsmouth City	WI	Racine
VA	Prince George	VA	Richmond City	WI	Sheboygan
VA	Prince William	VA	Roanoke City	WI	Waukesha
VA	Spotsylvania	VA	Suffolk	WY	Lincoln
VA	Stafford	VA	Virginia Beach	WY	Sublette
VA	York	VA	Williamsburg	WY	Sweetwater
VA	Alexandria	WA	Clark	PR	Bayamon
VA	Charlottesville	WA	King	PR	Mayaguez
VA	Chesapeake City	WA	Pierce	PR	San Juan

Appendix B
Sample Rebate Application – Public Equipment Owner

United States Environmental Protection Agency
National Clean Diesel Rebate Program
Rebate Application

OMB Number: 2060-0686
Expiration Date: 10/31/2015

Funding Year |2013|　Target Fleet |Construction|　Rebate Type |Retrofit/Replacement|

Applicant Information

Organization Name |Washtenaw County Road Commission|

Address |456 Easy Street|

City |Pleasantville|　County/Parish |Washtenaw|　State |MI|　ZIP |48108|

Employer/Taxpayer No. (EIN/TIN) |596001009|　Organizational DUNS Code |85440782|

Eligible Entity Information (Private Equipment Owner Applicants Only)
Private equipment owners are able to apply for funding from the National Clean Diesel Rebate Program if the equipment, for which funding is being requested, is currently contracted or leased to an eligible entity. An eligible entity is a federal, regional, State, local, or tribal agency or port authority with jurisdiction over transportation or air quality. For additional information regarding private equipment owner applicants and eligible entities, please refer to the Program Guide.

Eligible Entity Type	Eligible Entity Name	Eligible Entity Location (City, State)

☐ I certify the equipment, for which rebate funds are being requested, meet the requirements for private equipment owners as described above and in the terms and conditions within the Program Guide.

Original Equipment

	Equipment Type	Technology **	Engine Serial Number	Engine Family Name	Engine Model Year	Gals Used Last 12 Mos	Hrs of Use Last 12 Mos	HP	Location of Operation: County	State	Rebate Amount
*	Excavator	T1 to T3	8NC13041	YDPXL10.8MBF	2000	3759	500	225	Lake	IN	$15,000
*	Crane	T3 add DPF	MHX02842	6CPXL12.6ESK	2008	4632	800	388	Cook	IL	$30,000
1	Tractor/Loader/Backhoe	T1 to T3	C4E06383	8PKXL4.4NJ1	1996	6744	851	523	Washtenaw	MI	$69,000
2	Boring and Drilling Rig	T3 add DPF	44407183	9PKXL04.4NJ1	2010	5223	664	456	Washtenaw	MI	$30,000
3	Crane	UR to T2	RSX05086	6CPXL11.1ESK	2001	4985	6007	185	Washtenaw	MI	$12,000
4											
5											
*Example	** UR = unregulated, also referred to as Tier 0								Total		$ 111,000

☒ I certify that the engines listed for retrofit or replacement are operational and meet the eligibility requirements defined in the Program Guide.

☒ Replacements only: I certify that the engines listed for replacement will be properly disposed of according to the requirements defined in the Program Guide.

Applicant Signature

☒ By signing, I certify the statements and information provided in this application are true and accurate to the best of my knowledge. If selected for funding, I agree to provide the required documentation and assurances necessary for funding.

Funding for the National Clean Diesel Rebate Program is subject to continuing federal appropriations. Please see the Program Guide for additional funding information.

Authorized Representative Name	Jane Smith		
Title |Director| E-mail |jsmith@washtenawcountyroadcommission.gov| Phone |(734) 687-2584|

Authorized Representative Signature　*Jane Smith*　Date 1/5/2014

EPA Form 5600-260 (9-12)

Sample Rebate Application – Private Equipment Owner

EPA United States Environmental Protection Agency
National Clean Diesel Rebate Program
Rebate Application

OMB Number: 2060-0686
Expiration Date: 10/31/2015

Funding Year **2013**　Target Fleet **Construction**　Rebate Type **Retrofit/Replacement**

Applicant Information

Organization Name **Best Construction Company**

Address **123 Easy Street**

City **Pleasantville**　County/Parish **Washtenaw**　State **MI**　ZIP **48108**

Employer/Taxpayer No. (EIN/TIN) **38-2677401**　Organizational DUNS Code **926722823**

Eligible Entity Information (Private Equipment Owner Applicants Only)

Private equipment owners are able to apply for funding from the National Clean Diesel Rebate Program if the equipment, for which funding is being requested, is currently contracted or leased to an eligible entity. An eligible entity is a federal, regional, State, local, or tribal agency or port authority with jurisdiction over transportation or air quality. For additional information regarding private equipment owner applicants and eligible entities, please refer to the Program Guide.

Eligible Entity Type	Eligible Entity Name	Eligible Entity Location (City, State)
Local Govt/Agency	Washtenaw County Road Commission	Nice City, MI

[X] I certify the equipment, for which rebate funds are being requested, meet the requirements for private equipment owners as described above and in the terms and conditions within the Program Guide.

Original Equipment

	Equipment Type	Technology **	Engine Serial Number	Engine Family Name	Engine Model Year	Gals Used Last 12 Mos	Hrs of Use Last 12 Mos	HP	Location of Operation: County	State	Rebate Amount
*	Excavator	T1 to T3	6NC13641	YCPXL10.5MRF	2000	3759	500	225	Lake	IN	$15,000
*	Crane	T3 add DPF	MHX02642	6CPXL12.5ESK	2008	4632	600	368	Cook	IL	$30,000
1	Tractor/Loader/Backhoe	T1 to T3	C4E06383	8PKXL4.4NJ1	1996	6744	851	523	Washtenaw	MI	$69,000
2	Boring and Drilling Rig	T3 add DPF	44407183	9PKXL04.4NJ1	2010	5223	664	456	Washtenaw	MI	$30,000
3	Crane	UR to T3	RSX05086	6CPXL11.1ESK	2001	4985	6007	185	Washtenaw	MI	$15,000
4											
5											
*Example	** UR = unregulated, also referred to as Tier 0								Total		$114,000

[X] I certify that the engines listed for retrofit or replacement are operational and meet the eligibility requirements defined in the Program Guide.

[X] Replacements only: I certify that the engines listed for replacement will be properly disposed of according to the requirements defined in the Program Guide.

Applicant Signature

[X] By signing, I certify the statements and information provided in this application are true and accurate to the best of my knowledge. If selected for funding, I agree to provide the required documentation and assurances necessary for funding.

Funding for the National Clean Diesel Rebate Program is subject to continuing federal appropriations. Please see the Program Guide for additional information.

Authorized Representative Name	Lindsey Vitikainen		
Title Director of Equipment	E-mail lindseyv@bestconstructoincompany.com	Phone	(734) 687-2584

Authorized Representative Signature _Lindsey Vitikainen_　Date 1/5/2014

EPA Form 5600-260 (9-12)

Appendix C
Rebate Application Checklist

The following information is required in order for an application to be considered eligible and therefore entered into the random selection pool to potentially receive funds through the 2013 Construction Equipment Funding Opportunity.

EPA **must** receive the Rebate Application by **Wednesday, January 29, 2014, 4:00 pm EST.**

1) Review eligibility to apply (Section 2.1).

2) Review and determine equipment and engine eligibility including (sections 2.2, 2.3.1 and 2.3.2):

   a. Ownership
   b. Location Requirements
   c. Eligible Construction Equipment Engines
   d. Eligible Equipment Engine Model Year
   e. Eligible Equipment Engine Emission Standard Tier
   f. Eligible Equipment Engine Rated Power (Horsepower)

3) Certify Eligible Entity Information (Private Equipment Owner Applicants only).

4) Complete Rebate Application (EPA Form 5600-260) available for download at www.epa.gov/cleandiesel/dera-rebate-construction.htm

   a. Complete Equipment and Engine Information required on the rebate application for **each** engine to be retrofitted or replaced (up to five).

   b. Include Dun and Bradstreet (DUNS) and Employer ID Number (EIN).

   c. Provide name, title, contact information, and obtain the signature of the applicant's Authorized Representative.

5) Attach a scanned copy of the equipment's original bill of sale, invoice, or other ownership documentation for **each** engine to be retrofitted or replaced. Note: Equipment with an active lien-holder does not qualify to participate in this program.

6) Email the completed Rebate Application to: CleanDieselRebate@epa.gov Include in the subject line: DERA Construction Equipment Rebate Application: [your organization's name].

7) Selected applicants must be registered in the System for Award Management (SAM), (previously known as the Central Contractor Registration) prior to submitting an application. Information can be found at www.sam.gov.

Appendix D
How to Find Your Engine Serial Number, Engine Model Year, and Engine Family Name

D.1 Where to find the Engine Serial Number

The engine serial number is typically found in one of the following locations on Nonroad Engines:

1. Above air filter
2. Above the pulley, or on belt guard if present
3. Fuel pump or manifold
4. Engine block on the side of engine
5. On intake manifold
6. On/near starter
7. On cylinder head
8. On flange
9. On valve cover
10. Rear of engine block
11. On compressed air tank

The engine serial number may also be located on the original invoice or bill of lading. Detailed photographs of potential locations for the engine serial numbers can be found in Appendix H of the Construction Fleet Inventory Guide: www.epa.gov/cleandiesel/documents/420b10025.pdf

D.2 Where to find the Engine Model Year and Engine Family Name

The engine model year can be found on the label which is affixed to the engine itself. Do not be confused with the Equipment Model. The engine model year can differ from the equipment model year.

The EPA engine family name is an 11 or 12 character number/letter designation included on the engine nameplate for all nonroad engines sold in the United States. The engine family name is a 12-digit alpha-numeric code used by the U.S. EPA to classify vehicles and engines for the purpose of emissions certification. An engine may have an exhaust engine family name and an evaporative engine family name, depending on the year the engine was manufactured. The engine family name can also be located on a label or plate in the engine compartment in a visible position such as the hood underside, shock tower, radiator support, fan shroud, or firewall.

If the engine label does not list the EPA-issued engine family name or is difficult to read, the manufacturer of your engine may be able to assist you in determining the engine family name if you are able to supply information on the model year and make of the engine, or the engine serial number. Some manufacturers also have online tools where their customer can enter engine serial numbers and determine the engine family name.

Detailed photographs of potential locations where the Family Engine name may be found are in Appendix H of the Construction Fleet Inventory Guide. The Construction Fleet Inventory Guide is available at:
www.epa.gov/cleandiesel/documents/420b10025.pdf

Appendix E
Technology Option # 1 Eligibility and Rebate Amount Worksheet
Retrofit with a Diesel Particulate Filter

1. Equipment Type: _____

2. Fill in the blanks below, then use the table to determine current engine emission standard tier level. Please note that engine model year may differ from equipment model year.

    Horsepower: _____
    Engine Model Year: _____
    Enter Current Engine
    Emission Standard Tier: _____

Current Engine Horsepower	Current Engine Model Year	Current Emission Standard Tier
174-300	2003-2005	Tier 2
174-300	2006-2010	Tier 3
301- 603	2001-2005	Tier 2
301- 603	2006-2010	Tier 3

3. Use the below table to determine Rebate Amount

Current Emission Standard Tier	Current Engine Horsepower	Rebate Amount
Tier 2	174-300	Not Eligible
Tier 3	174-300	Not Eligible
Tier 2	301- 603	$30,000
Tier 3	301- 603	$30,000

    Rebate Amount: _____

4. Select the above amount in the drop-down Rebate Amount Box on the Rebate Application.

This worksheet does not need to be submitted with the Rebate Application.

Appendix F
Technology Option # 2 Eligibility and Rebate Amount Worksheet
Engine Replacement

1. Equipment Type: _____

2. Fill in the blanks below, then use the table to determine current engine emission standard tier level. Please note that engine model year may differ from equipment model year.

    Horsepower: _____
    Engine Model Year: _____
    Enter Current Engine
    Emission Standard Tier: _____

Current Engine HP	Current Engine Model Year	Current Tier
174-300	1990-1995	Unregulated
174-300	1996-2002	Tier 1
301- 603	1990-1995	Unregulated
301- 603	1996-2000	Tier 1

3. Circle the Replacement Engine Tier
    Tier 2   Tier 3

4. Select the Replacement Tier and Determine Rebate Amount

Current Tier	Replacement Tier	Rebate Amount
Unregulated (UR)	Tier 2 (174-300hp)	$12,000
Unregulated (UR)	Tier 3 (174-300hp)	$15,000
Tier 1	Tier 2 (174-300hp)	$12,000
Tier 1	Tier 3 (174-300hp)	$15,000
Unregulated (UR)	Tier 2 (301- 603hp)	$49,000
Unregulated (UR)	Tier 3 (301- 603hp)	$69,000
Tier 1	Tier 2 (301- 603hp)	$49,000
Tier 1	Tier 3 (301- 603hp)	$69,000

Rebate Amount $_____

5. Select the above amount in the drop-down Rebate Amount Box on the Rebate Application.

    This worksheet does not need to be submitted with the Rebate Application.

www.ingramcontent.com/pod-product-compliance
Lightning Source LLC
Chambersburg PA
CBHW081413170526
45166CB00010B/3328